THE POETRY OF SAMARIUM

The Poetry of Samarium

Walter the Educator

Silent King Books a WhichHead Imprint

Copyright © 2023 by Walter the Educator

All rights reserved. No part of this book may be reproduced in any manner whatsoever without written permission except in the case of brief quotations embodied in critical articles and reviews.

First Printing, 2023

Disclaimer
This book is a literary work; poems are not about specific persons, locations, situations, and/or circumstances unless mentioned in a historical context. This book is for entertainment and informational purposes only. The author and publisher offer this information without warranties expressed or implied. No matter the grounds, neither the author nor the publisher will be accountable for any losses, injuries, or other damages caused by the reader's use of this book. The use of this book acknowledges an understanding and acceptance of this disclaimer.

"Earning a degree in chemistry changed my life!"
- Walter the Educator

dedicated to all the chemistry lovers, like myself, across the world

CONTENTS

Dedication v

Why I Created This Book? 1

One - Number 62 2

Two - Samarium, Oh Samarium 4

Three - Forever Divine 6

Four - Captivating World 8

Five - Element We Adore 10

Six - Celestial Bond 12

Seven - Cosmic Companion 14

Eight - Atomic Symphony 16

Nine - Celebrate Samarium 18

Ten - Binds All Things 20

Eleven - The Universe Sings 22

Twelve - Makes Us Believe 24

Thirteen - Samarium, The Element	. . . . . .	26
Fourteen - Secrets To Uncover	. . . .	28
Fifteen - Bridge Between Worlds	. . . . . .	30
Sixteen - Harmonizing Lives	. . . .	32
Seventeen - Illuminating The Dark	. . . . . .	34
Eighteen - Samarium's Essence	. . . .	36
Nineteen - Burning Brighter	. . . .	38
Twenty - Stellar Muse	. . . .	40
Twenty-One - Through The Night	. . . .	42
Twenty-Two - Dance And Twirl	. . . .	44
Twenty-Three - Cherish The Magic	. . . . . .	46
Twenty-Four - Weaves Stories Untold	. . . .	48
Twenty-Five - Dreams Unfurl	. . . .	50
Twenty-Six - Spirits Wings	. . . .	52
Twenty-Seven - Earthly Things	. . . .	54
Twenty-Eight - Harmony Brings	. . . .	56
Twenty-Nine - Enchants The World	. . . . . .	58
Thirty - Lead Into Gold	. . . .	60
Thirty-One - Horizons Take Flight	. . . . . .	62
Thirty-Two - With Samarium	. . . .	64

Thirty-Three - Samarium As Our Guide . . . 66

Thirty-Four - Samarium's Touch 68

About The Author 70

WHY I CREATED THIS BOOK?

Creating a poetry book about the chemical element of Samarium was a unique and fascinating endeavor. Samarium, a rare earth metal, possesses intriguing properties that serve as inspiration for poetic exploration. Its atomic structure, physical characteristics, and historical significance can be woven into verses, allowing for a fusion of science and art. This project can educate readers about Samarium while engaging them with the beauty of language and imagery. By delving into the element's properties, symbolism, and potential applications, this poetry book can illuminate the wonders of Samarium in a creative and thought-provoking manner.

ONE

NUMBER 62

In the realm of elements, let me speak of Samarium,
A treasure from the periodic table, a shimmering phenomenon.
With atomic number 62, it holds a mystical grace,
A tale of wonders hidden within its atomic space.

Samarium, oh Samarium, a rare and noble gem,
Found in minerals and ores, a celestial diadem.
A lustrous metal, glowing in shades of silver and blue,
Its magnetic allure capturing hearts, both old and new.

Within the Earth's crust, Samarium claims its abode,
A silent guardian, a secret it has bestowed.
With properties unique, it dances with electrons in its shell,

A symphony of bonds, a chemistry that no one can tell.

In magnets and alloys, Samarium finds its might,
Enhancing strength and power, a beacon in the night.
A catalyst of change, it sparks reactions divine,
Unleashing energy, the universe's grand design.

But beyond its scientific worth, a deeper tale unfolds,
Of Samarium's essence, of stories yet untold.
For in the vast cosmos, it whispers ancient lore,
A connection to the stars, a mystery to explore.

So let us celebrate Samarium, this element of awe,
A symbol of discovery, a testament to nature's law.
In its atomic dance, a symphony of cosmic art,
Samarium, a marvel that ignites the human heart.

TWO

SAMARIUM, OH SAMARIUM

In the realm of elements, there lies Samarium,
A cosmic spark, a luminescent anthem.
With atomic number 62, it holds its own sway,
A story untold, unfolding each passing day.
    Samarium, oh Samarium, a beacon of rare light,
A treasure concealed, shining ever so bright.
In nature's embrace, it finds its earthly home,
A gem of secrets, waiting to be known.
    Within its core, a magnetic symphony plays,
Dancing with electrons in mystical arrays.
A conductor of energy, it harmonizes the scene,
Unveiling mysteries, where science and wonder convene.
    From ancient lands, it emerges with grace,

A silent guardian, entwined in time's embrace.
Its hues of silver and blue, a celestial art,
A cosmic reminder of the universe's start.

    Samarium, a catalyst of change, a catalyst of dreams,
Unleashing potential with vibrant beams.
In magnets and lasers, its power unfurls,
A force of transformation, that enchants the world.

    But beyond its science, lies a deeper tale,
An alchemist's secret, that will never fail.
Samarium, oh Samarium, a symbol of grace,
A celestial element, in this boundless space.

THREE

FOREVER DIVINE

In the realm of elements, behold Samarium's might,
A luminous jewel, casting a celestial light.
With atomic number 62, it claims its domain,
A magnetic enigma, defying the mundane.

Samarium, oh Samarium, a treasure yet unknown,
A symphony of wonders in its atomic throne.
Within the Earth's embrace, it quietly resides,
A secret guardian, where wisdom abides.

Its metal form, a dance of shades silver and blue,
Magnetic allure, captivating all those who pursue.
A beacon of strength, in magnets it thrives,
Harnessing power, where possibilities derive.

But Samarium's tale extends beyond the realm of science,
Unveiling a narrative that sparks deep reliance.

For within its essence, a connection to the stars,
A cosmic tapestry woven, transcending our memoirs.
　In the vast expanse of space, it whispers ancient songs,
Echoing the mysteries, where the universe belongs.
Samarium, oh Samarium, a bridge to the unknown,
A celestial storyteller, in elements it's shown.
　So let us celebrate Samarium, this element of grace,
An enigmatic force, painting our world's embrace.
In its magnetic symphony, secrets intertwine,
Samarium, a cosmic rhapsody, forever divine.

FOUR

CAPTIVATING WORLD

In the realm of elements, Samarium prevails,
A mystical presence, where enchantment trails.
With atomic number 62, it reigns supreme,
A treasure untamed, a luminescent dream.
    Samarium, oh Samarium, a star in disguise,
A cosmic magician, captivating our eyes.
Within the Earth's core, it finds its hidden place,
A guardian of secrets, a celestial embrace.
    In shades of silver and blue, it radiates grace,
A touch of elegance, lighting up space.
Its magnetic prowess, a force to behold,
Drawing hearts closer, as stories unfold.
    Beyond the realm of science, Samarium weaves,
A tapestry of wonder, where mystery conceives.

With each electron dance, a symphony of delight,
Unveiling the universe, in colors so bright.
   In magnets and lasers, its power takes flight,
Igniting innovation, pushing boundaries with might.
A catalyst for change, a catalyst for the soul,
Samarium, the alchemist's ultimate goal.
   So let us celebrate Samarium, this elemental star,
A cosmic luminary, both near and far.
In its atomic dance, secrets are unfurled,
Samarium, the embodiment of a captivating world.

FIVE

ELEMENT WE ADORE

In the realm of elements, a hidden gem resides,
Samarium, the enigmatic, where secrets coincide.
A dancer of electrons, in orbits it twirls,
Unveiling its wonders, as the universe unfurls.

With a glow of silver and blue, it captivates the eye,
A celestial jewel, shining bright in the sky.
Magnetic in essence, it pulls and repels,
A force of attraction, where science compels.

But beyond its scientific allure, a deeper tale unfolds,
Of Samarium's essence, where cosmic mysteries enfold.
For in the cosmic orchestra, it plays a vital part,
Connecting the dots, weaving stars and hearts.

A catalyst of change, it sparks fires within,

Igniting innovation, where possibilities begin.
In lasers and magnets, its power takes flight,
Unleashing potential, pushing boundaries with might.

So let us celebrate Samarium, this elemental grace,
A cosmic storyteller, in the vast cosmic space.
In its atomic dance, it whispers ancient lore,
Samarium, the cosmic element we adore.

SIX

CELESTIAL BOND

In the realm of elements, a starry delight,
Samarium emerges, casting a vibrant light.
With atomic artistry, it dances and spins,
A cosmic conductor, where new stories begin.
    Magnetic in nature, it draws near and far,
A symphony of forces, guiding us like a star.
In magnets and lasers, its power finds a stage,
Unleashing potential, pushing boundaries with rage.
    But beyond its science, a tale yet untold,
Samarium's secrets, in mysteries unfold.
In distant galaxies, its essence takes flight,
A celestial bond, connecting day and night.
    So let us honor Samarium, this cosmic guide,
A bridge between worlds, where wonders reside.

In its atomic tapestry, secrets are unfurled,
Samarium, the catalyst of a cosmic world.

SEVEN

COSMIC COMPANION

In the realm of elements, Samarium holds its reign,
A cosmic jewel, with a captivating refrain.
Its atomic heartbeat, a rhythm of its own,
Unveiling secrets, in a universe unknown.

Magnetic in nature, it draws forces together,
Creating bonds, like an eternal tether.
In magnets and lasers, its power takes flight,
Igniting innovation, with brilliant light.

But beyond its science, a deeper tale lies,
Where Samarium dances, with celestial ties.
In distant galaxies, its presence is felt,
A guiding star, where mysteries are knelt.

Samarium, oh Samarium, a cosmic symphony,
Playing the chords of infinity.

In its atomic dance, secrets are unfurled,
A cosmic conductor, in a boundless world.
    So let us embrace Samarium, this elemental gem,
A cosmic companion, beyond what we condemn.
In its atomic embrace, wonders come alive,
Samarium, the celestial element, forever to thrive.

EIGHT

ATOMIC SYMPHONY

In the realm of elements, Samarium stands tall,
A dancer of electrons, captivating us all.
With its atomic grace, it takes center stage,
Unveiling cosmic secrets, like an ancient sage.

Magnetism its prowess, a force to admire,
Attracting and repelling, like a celestial choir.
In magnets and lasers, its power takes flight,
Harnessing energy, pushing boundaries with might.

But beyond its scientific allure and might,
Samarium holds secrets, shining in cosmic light.
In distant galaxies, its presence is known,
A cosmic storyteller, with tales yet to be shown.

Samarium, oh Samarium, a celestial guide,
A bridge between worlds, where wonders coincide.

In its atomic dance, mysteries intertwine,
Connecting the universe, in a grand design.

So let us celebrate Samarium, this element of awe,
A cosmic orchestra, playing nature's sweetest score.
In its atomic symphony, secrets are unfurled,
Samarium, the cosmic alchemist, transforming the world.

NINE

CELEBRATE SAMARIUM

In the realm of elements, a gem doth shine,
Samarium, a cosmic treasure, so divine.
With atomic elegance, it takes its place,
Embracing the universe with its grace.
   Magnetic by nature, it holds a special might,
Drawing energies together, shining bright.
In magnets and lasers, its power takes flight,
Unleashing innovation, breaking through the night.
   But beyond its science, a deeper story unfolds,
Samarium, the cosmic weaver, it beholds.
In distant galaxies, it paints a cosmic scene,
Connecting the stars, where dreams convene.
   Samarium, oh Samarium, a celestial guide,
A bridge between worlds, where mysteries reside.

In its atomic dance, the universe comes alive,
A cosmic conductor, harmonizing the sky.
   So let us celebrate Samarium, this elemental star,
A cosmic navigator, guiding us from afar.
In its atomic symphony, secrets are unfurled,
Samarium, the cosmic maestro, transforming our world.

TEN

BINDS ALL THINGS

Samarium, the element of celestial grace,
In the periodic table, it finds its rightful place.
With atomic allure, it captures the eye,
A cosmic dancer, painting the sky.

Magnetic and rare, its essence enchants,
In magnets and lasers, its power enhances.
But beyond the science, a deeper tale unfolds,
Samarium's secrets, the universe holds.

In distant galaxies, it weaves a cosmic thread,
Connecting the stars, where mysteries spread.
A celestial guide, in the vast cosmic sea,
Samarium, the element that sets us free.

Oh Samarium, a catalyst of change,
Innovation and progress, in its atomic range.
With every interaction, a story is told,
Of cosmic connections and wonders untold.

So let us embrace Samarium, this cosmic jewel,
A bridge between worlds, where dreams fuel.
In its atomic symphony, the universe sings,
Samarium, the element that binds all things.

ELEVEN

THE UNIVERSE SINGS

In the realm of elements, Samarium shines,
A cosmic dancer, with secrets it aligns.
From magnets to lasers, its power extends,
A catalyst of change, where innovation transcends.

But beyond its science, a deeper story unfolds,
Samarium, the element, its wonders yet untold.
In celestial symphony, it plays a part,
Connecting the pieces, weaving the cosmic art.

In distant galaxies, its presence is felt,
A celestial navigator, where mysteries are dwelt.
Guiding the stars, with an ethereal grace,
Samarium, the cosmic compass in space.

Oh Samarium, a conductor of cosmic cheer,
With atomic melodies that reach far and near.

In its atomic dance, it paints celestial hues,
A cosmic poet, crafting stories infused.
  Let us celebrate Samarium, this element divine,
A bridge between worlds, where secrets entwine.
In its atomic symphony, the universe sings,
Samarium, the cosmic element that brings.

TWELVE

MAKES US BELIEVE

Samarium, a jewel among the stars,
With atomic elegance, it leaves its mark.
In magnets and lasers, its power ignites,
A cosmic fire, illuminating the nights.

But beyond its science, a tale yet to unfold,
Samarium's secrets, waiting to be told.
In distant galaxies, its essence takes flight,
A celestial bond, connecting day and night.

Oh Samarium, a cosmic guide so rare,
Navigating the cosmos with utmost care.
In its atomic dance, wonders come alive,
A celestial conductor, harmonizing the skies.

Let us honor Samarium, this elemental gem,
A bridge between worlds, a cosmic anthem.
In its atomic embrace, mysteries unfurl,

Samarium, the cosmic alchemist, transforming the world.

So let us marvel at Samarium's cosmic grace,
Unveiling the wonders of time and space.
In its atomic symphony, a cosmic story is weaved,
Samarium, the element that makes us believe.

THIRTEEN

SAMARIUM, THE ELEMENT

Samarium, a marvel in the atomic realm,
A cosmic dancer, wearing a stellar helm.
In magnets and lasers, its power takes flight,
Igniting innovation, casting celestial light.

Beyond the science, a deeper tale unfurls,
Samarium, the cosmic conductor of swirling worlds.
In distant galaxies, its presence does transcend,
Connecting the cosmos, where mysteries blend.

Oh Samarium, a celestial architect so grand,
Building bridges 'twixt stars, as the universe expands.
In its atomic ballet, secrets are revealed,
A cosmic symphony, where wonders are concealed.

Let us celebrate Samarium, this element of awe,
A cosmic navigator, defying nature's laws.

In its atomic embrace, the universe is aligned,
Samarium, the cosmic alchemist, forever enshrined.

So let us ponder Samarium's cosmic role,
A stardust conductor, guiding every celestial stroll.
In its atomic symphony, a cosmic story is unveiled,
Samarium, the element that keeps the cosmos compelled.

FOURTEEN

SECRETS TO UNCOVER

Samarium, a shimmering cosmic gem,
With atomic secrets, a celestial diadem.
In distant galaxies, its presence gleams,
Connecting the cosmos in ethereal streams.
    Oh Samarium, a celestial lighthouse so bright,
Guiding the stars through the vast expanse of night.
In its atomic dance, mysteries intertwine,
A cosmic weaver, creating a tapestry divine.
    Let us rejoice in Samarium's cosmic embrace,
A bridge between realms, traversing time and space.
In its atomic symphony, the universe sings,
Samarium, the catalyst that harmony brings.
    So let us marvel at Samarium's cosmic might,
A celestial magician, transforming darkness to light.

In its atomic ballet, the cosmos comes alive,
Samarium, the elixir through which dreams thrive.

Embrace Samarium, this element of cosmic wonder,
A celestial storyteller, with secrets to uncover.
In its atomic orchestra, the universe it molds,
Samarium, the cosmic key, unlocking mysteries untold.

FIFTEEN

BRIDGE BETWEEN WORLDS

Samarium, a cosmic gem, shining bright,
In the vast expanse of celestial night.
With atomic grace, it dances and glows,
A stellar performer, the cosmos bestows.

In distant galaxies, its secrets unfurl,
Connecting the stars, a cosmic whirl.
Samarium, the catalyst of cosmic fire,
Igniting the universe with its atomic desire.

Oh Samarium, a guide through the astral plane,
Navigating the cosmos, its purpose arcane.
In its atomic symphony, galaxies align,
Samarium, the cosmic architect, so divine.

Let us celebrate Samarium's cosmic role,
A bridge between worlds, the universe extol.

In its atomic embrace, wonders are revealed,
Samarium, the cosmic alchemist, forever sealed.

    Embrace Samarium's essence, a celestial prize,
Unveiling the mysteries that lie within the skies.
In its atomic ballet, the universe sings,
Samarium, the element that harmony brings.

SIXTEEN

HARMONIZING LIVES

Samarium, the element of cosmic might,
A celestial dancer in the cosmic night.
In distant galaxies, its presence shines,
Connecting the stars with radiant lines.
    Oh Samarium, a celestial guide so rare,
Navigating the cosmos with utmost care.
In its atomic dance, the universe thrives,
A cosmic conductor, harmonizing lives.
    Let us honor Samarium, this cosmic gem,
A bridge between worlds, a celestial anthem.
In its atomic embrace, the cosmos unfolds,
Samarium, the cosmic alchemist, secrets it holds.
    So let us marvel at Samarium's cosmic grace,
Unveiling the wonders of time and space.

In its atomic symphony, a cosmic story is told,
Samarium, the element that makes the universe bold.
  Embrace Samarium's essence, a cosmic key,
Unlocking the mysteries of eternity.
In its atomic ballet, celestial dreams ignite,
Samarium, the element that guides our cosmic flight.

SEVENTEEN

ILLUMINATING THE DARK

Samarium, a cosmic jewel, shining bright,
In the cosmic dance, it weaves its light.
A bridge between realms, it transcends,
Connecting the stars, where wonder never ends.
    Oh Samarium, a celestial conductor, so rare,
Guiding the universe with cosmic flair.
In its atomic symphony, galaxies align,
Samarium, the cosmic maestro, so divine.
    Let us celebrate Samarium, this element of awe,
A cosmic navigator, defying nature's law.
In its atomic embrace, the universe is transformed,
Samarium, the alchemist, through which wonders are formed.
    Embrace Samarium's essence, a cosmic spark,

Igniting the cosmos, illuminating the dark.
In its atomic ballet, celestial stories unfold,
Samarium, the element that makes the universe bold.

So let us marvel at Samarium's cosmic might,
A stellar luminary, casting radiant light.
In its atomic symphony, the universe sings,
Samarium, the cosmic jewel that forever brings.

EIGHTEEN

SAMARIUM'S ESSENCE

Samarium, the element of cosmic might,
In the celestial realm, it shines so bright.
A beacon in the depths of the unknown,
Guiding us through the universe, it's shown.
   Oh Samarium, a cosmic conductor grand,
Orchestrating the dance of atoms in every strand.
In its atomic symphony, the stars align,
Samarium, the cosmic architect, so fine.
   Let us celebrate the wonders it imparts,
Unveiling the secrets of the cosmos, stealing hearts.
In its atomic embrace, galaxies unfold,
Samarium, the cosmic alchemist, turning stories gold.
   Embrace Samarium's essence, a cosmic key,
Unlocking the mysteries of eternity.

In its atomic ballet, celestial dreams ignite,
Samarium, the element that guides our cosmic flight.
   So let us marvel at Samarium's cosmic grace,
A celestial traveler, exploring time and space.
In its atomic symphony, the universe sings,
Samarium, the cosmic gem that forever brings.

NINETEEN

BURNING BRIGHTER

In the realm of elements, a star is born,
Samarium, a cosmic jewel adorned.
Its atomic rhythm, a celestial beat,
Guiding the universe with grace and fleet.
    Oh Samarium, a conductor of light,
Weaving constellations, shimmering bright.
In its atomic dance, galaxies unite,
Samarium, the cosmic catalyst of infinite.
    Let us celebrate this element divine,
A bridge between worlds, where mysteries align.
In its atomic embrace, secrets unfurl,
Samarium, the cosmic alchemist, a precious pearl.
    Embrace the essence of Samarium's might,
A celestial navigator, leading us through the night.

In its atomic symphony, harmony rings,
Samarium, the cosmic muse that forever sings.
    So let us marvel at Samarium's cosmic power,
A cosmic flame, burning brighter by the hour.
In its atomic ballet, the universe takes flight,
Samarium, the element that ignites our cosmic sight.

TWENTY

STELLAR MUSE

In the realm of atoms, a celestial dancer,
Samarium twirls, a cosmic enhancer.
In its atomic symphony, elegance prevails,
A cosmic conductor, weaving cosmic tales.
    Let us celebrate Samarium, this cosmic gem,
A bridge between worlds, a celestial emblem.
In its atomic embrace, wonders come alive,
Samarium, the cosmic alchemist, where dreams thrive.
    Embrace the essence of Samarium's embrace,
A cosmic navigator, exploring time and space.
In its atomic ballet, the universe unfolds,
Samarium, the cosmic architect, secrets it holds.
    So let us marvel at Samarium's cosmic art,
A stellar muse, igniting passion in every heart.

In its atomic symphony, the stars align,
Samarium, the element that makes the cosmos shine.
   Embrace Samarium's essence, a celestial gift,
Unlocking the mysteries that the universe lifts.
In its atomic grace, celestial wonders ignite,
Samarium, the element that guides our cosmic flight.

TWENTY-ONE

THROUGH THE NIGHT

In the cosmic tapestry, a gleaming thread,
Samarium, the celestial jewel, it's said.
In its atomic dance, the universe aligns,
A cosmic conductor, orchestrating divine signs.
Let us celebrate Samarium's cosmic reign,
A bridge between realms, where galaxies remain.
In its atomic embrace, secrets are revealed,
Samarium, the cosmic alchemist, forever sealed.
Embrace the essence of Samarium's might,
A stellar beacon, casting celestial light.
In its atomic ballet, cosmic stories unfold,
Samarium, the element that turns legends bold.
So let us marvel at Samarium's cosmic glow,
A celestial traveler, through time it does flow.
In its atomic symphony, a cosmic narrative rings,
Samarium, the element that cosmic wonders bring.

Embrace Samarium's essence, a celestial key,
Unlocking the mysteries of the vast cosmic sea.
In its atomic dance, the universe takes flight,
Samarium, the element that guides us through the night.

TWENTY-TWO

DANCE AND TWIRL

In the realm of atoms, a cosmic luminary,
Samarium, the element that paints the celestial diary.
In its atomic embrace, cosmic tapestries unfurl,
Samarium, the alchemist, crafting a celestial swirl.

Embrace the essence of Samarium's cosmic might,
A stellar conductor, orchestrating cosmic light.
In its atomic ballet, galaxies dance and twirl,
Samarium, the cosmic architect, creating a cosmic pearl.

So let us marvel at Samarium's celestial grace,
A cosmic wanderer, traversing time and space.
In its atomic symphony, celestial stories are told,
Samarium, the element that makes the cosmos bold.

Embrace Samarium's essence, a celestial guide,
Unlocking the secrets of the universe far and wide.

In its atomic dance, cosmic wonders ignite,
Samarium, the element that sets our souls alight.

TWENTY-THREE

CHERISH THE MAGIC

In the realm of atoms, a cosmic jewel,
Samarium, the element, wise and cool.
In its atomic embrace, secrets unfold,
Samarium, the cosmic alchemist, stories untold.

    Embrace the essence of Samarium's might,
A celestial navigator, shining bright.
In its atomic ballet, cosmic symphonies play,
Samarium, the element that lights up the way.

    Let us marvel at Samarium's cosmic art,
A stellar architect, sculpting every part.
In its atomic dance, the universe gleams,
Samarium, the cosmic thread that weaves dreams.

    Embrace Samarium's essence, a cosmic key,
Unlocking the wonders of eternity.
In its atomic symphony, celestial tales ignite,

Samarium, the element that guides us through the night.

    So let us cherish the magic Samarium weaves,
A cosmic journey that our imagination receives.
In its atomic embrace, the universe sings,
Samarium, the cosmic gem that forever brings.

TWENTY-FOUR

WEAVES STORIES UNTOLD

In the depths of cosmic tapestry,
Samarium, a luminary entity.
In its atomic embrace, secrets unfurl,
A cosmic alchemist, a celestial pearl.
    Embrace the essence of Samarium's might,
A stellar navigator, guiding us through the night.
In its atomic ballet, constellations align,
Samarium, the cosmic maestro, melodies divine.
    Let us marvel at Samarium's cosmic fire,
An element that sparks celestial desire.
In its atomic symphony, stars are born,
Samarium, the cosmic catalyst, forever adorned.
    So let us embrace Samarium's cosmic glow,
A celestial wanderer, with secrets to bestow.

In its atomic dance, galaxies unfold,
Samarium, the element that weaves stories untold.
  Embrace Samarium's essence, a cosmic key,
Unlocking the wonders of the universe, for all to see.
In its atomic embrace, the cosmos takes flight,
Samarium, the element that fills us with cosmic light.

TWENTY-FIVE

DREAMS UNFURL

In the celestial realm, a mystery untold,
Samarium, the cosmic gem, radiant and bold.
In its atomic dance, a cosmic enchantress,
Samarium, the element that brings cosmic finesse.

Let us marvel at Samarium's celestial sway,
A stellar conductor, guiding the astral display.
In its atomic symphony, the universe sings,
Samarium, the cosmic essence that harmony brings.

Embrace Samarium's essence, a cosmic key,
Unlocking the secrets of the cosmic sea.
In its atomic embrace, celestial wonders ignite,
Samarium, the element that paints the cosmic night.

So let us cherish Samarium's cosmic embrace,
A celestial traveler, traversing time and space.
In its atomic dance, cosmic stories unfold,
Samarium, the element that transcends the mold.

Embrace Samarium's essence, a celestial guide,
Unlocking the cosmic wisdom that resides.
In its atomic ballet, cosmic dreams unfurl,
Samarium, the element that lights up the world.

TWENTY-SIX

SPIRITS WINGS

Samarium, the silent cosmic dancer,
Whispering secrets in the celestial expanse.
In its atomic waltz, the universe glows,
A symphony of stars, it gracefully shows.

A stellar chameleon, Samarium's hue,
Shifting and changing, revealing something new.
In its atomic tapestry, colors intertwine,
Creating cosmic artwork, divine and fine.

Embrace Samarium's essence, a cosmic spell,
Unveiling the wonders that no words can tell.
In its atomic symphony, wonders unfold,
Samarium, the element that never grows old.

Let us marvel at Samarium's cosmic flight,
A celestial navigator, guiding us through the night.
In its atomic dance, galaxies align,
Samarium, the cosmic compass, serene and benign.

Embrace Samarium's essence, a celestial key,
Unlocking the mysteries of the vast cosmic sea.
In its atomic embrace, dreams take flight,
Samarium, the element that ignites cosmic light.

   So let us cherish Samarium's cosmic glow,
A celestial journey, where wonders forever flow.
In its atomic ballet, the universe sings,
Samarium, the element that gives our spirits wings.

TWENTY-SEVEN

EARTHLY THINGS

In the depths of the universe, a cosmic gem,
Samarium, the celestial diadem.
Its atomic dance, a mesmerizing display,
A stellar symphony that lights up the Milky Way.
    Embrace Samarium's essence, a cosmic key,
A window to the cosmos, for all to see.
In its atomic embrace, the stars align,
Samarium, the element that makes the galaxies shine.
    Let us marvel at Samarium's cosmic grace,
A celestial guide in the vast cosmic space.
In its atomic ballet, constellations twirl,
Samarium, the cosmic conductor, unfurls.
    Embrace Samarium's essence, a stellar flame,
A cosmic luminary with an eternal name.

In its atomic symphony, celestial dreams ignite,
Samarium, the element that brings cosmic delight.
   So let us cherish Samarium's cosmic might,
A celestial wanderer, glowing in the night.
In its atomic embrace, the universe sings,
Samarium, the element that transcends earthly things.

TWENTY-EIGHT

HARMONY BRINGS

In the realm of atoms, a cosmic star,
Samarium shines, both near and far.
Its electrons dance, in a celestial waltz,
A symphony of cosmic pulses, it exalts.

A magnetic allure, Samarium possesses,
Guiding us through the universe's recesses.
In its atomic embrace, fields intertwine,
Samarium, the element, magnetic and divine.

It paints the cosmos with colors unknown,
A cosmic artist, its brilliance shown.
In its atomic ballet, celestial hues ignite,
Samarium, the element that colors the night.

Embrace Samarium's essence, a celestial key,
Unlocking the mysteries that lie beyond what we see.

In its atomic dance, secrets unfurl,
Samarium, the element that shapes the world.
    So let us cherish Samarium's cosmic reign,
A celestial ruler, where wonders remain.
In its atomic embrace, the universe sings,
Samarium, the element that cosmic harmony brings.

TWENTY-NINE

ENCHANTS THE WORLD

In the realm of elements, Samarium shines,
A cosmic gem with celestial designs.
In its atomic symphony, beauty takes flight,
Samarium, the star that illuminates the night.

Embrace Samarium's essence, a cosmic force,
An element that steers a celestial course.
In its atomic embrace, mysteries unfold,
Samarium, the catalyst of stories untold.

It dances with stardust, a celestial waltz,
Samarium, the element that never exhausts.
In its atomic ballet, galaxies align,
Samarium, the cosmic conductor, so fine.

Let us marvel at Samarium's cosmic fire,
An element that sparks celestial desire.

In its atomic embrace, dreams take flight,
Samarium, the element that ignites cosmic light.

So let us cherish Samarium's cosmic embrace,
A celestial traveler, exploring time and space.
In its atomic dance, cosmic wonders unfurl,
Samarium, the element that enchants the world.

Embrace Samarium's essence, a cosmic guide,
Unlocking the mysteries that the universe hides.
In its atomic symphony, the cosmos sings,
Samarium, the element that gives us cosmic wings.

THIRTY

LEAD INTO GOLD

In the realm of elements, Samarium stands tall,
A cosmic warrior, enchanting us all.
In its atomic dance, secrets it reveals,
Samarium, the element that cosmic fate seals.
    Embrace Samarium's essence, a celestial gift,
Unlocking the mysteries that the universe sift.
In its atomic embrace, stars come alive,
Samarium, the element that makes us thrive.
    Let us marvel at Samarium's cosmic might,
A celestial beacon, guiding us through the night.
In its atomic symphony, galaxies align,
Samarium, the element that makes the cosmos shine.
    So let us cherish Samarium's cosmic spell,
A celestial alchemist, weaving wonders so well.

In its atomic ballet, dreams are spun,
Samarium, the element that connects everyone.
   Embrace Samarium's essence, a cosmic key,
Unlocking the doors to the unknown, you see.
In its atomic embrace, cosmic stories unfold,
Samarium, the element that turns lead into gold.

THIRTY-ONE

HORIZONS TAKE FLIGHT

In the realm of elements, Samarium stands tall,
A cosmic treasure, captivating us all.
In its atomic dance, a symphony of grace,
Samarium, the element that paints the cosmic space.

Let us delve into Samarium's cosmic lore,
A celestial traveler, forever exploring more.
In its atomic embrace, secrets come alive,
Samarium, the element that helps dreams thrive.

Embrace Samarium's essence, a celestial guide,
Unlocking the mysteries where the universe hides.
In its atomic ballet, new horizons take flight,
Samarium, the element that ignites cosmic light.

So let us cherish Samarium's cosmic reign,
A celestial beacon, amidst the cosmic terrain.

In its atomic embrace, cosmic wonders unfold,
Samarium, the element that turns stories bold.
  With Samarium, the cosmos becomes our muse,
A cosmic canvas where imagination can amuse.
In its atomic symphony, the stars align,
Samarium, the element that makes the universe shine.

THIRTY-TWO

WITH SAMARIUM

In the realm of elements, a cosmic gem we find,
Samarium, a celestial treasure of a unique kind.
In its atomic dance, a cosmic story unfurls,
Samarium, the element that enchants and swirls.
    Let us marvel at Samarium's celestial grace,
A cosmic conductor, guiding stars in their chase.
In its atomic embrace, constellations align,
Samarium, the element that makes galaxies shine.
    Embrace Samarium's essence, a celestial key,
Unveiling the mysteries of the vast cosmic sea.
In its atomic symphony, cosmic dreams ignite,
Samarium, the element that paints cosmic light.
    So let us cherish Samarium's cosmic reign,
A celestial navigator, a guide we can't explain.

In its atomic embrace, the universe sings,
Samarium, the element that spreads cosmic wings.
    With Samarium, the cosmos becomes our muse,
A cosmic tapestry where wonder and beauty fuse.
In its atomic ballet, let our spirits take flight,
Samarium, the element that ignites cosmic delight.

THIRTY-THREE

SAMARIUM AS OUR GUIDE

In the realm of atoms, where wonders reside,
Samarium emerges, a cosmic guide.
In its atomic dance, a symphony unfolds,
Samarium, the element that cosmic beauty beholds.
 Embrace Samarium's essence, a celestial spark,
Igniting the night with a mystical arc.
In its atomic embrace, stars come alive,
Samarium, the element that helps dreams thrive.
 Let us marvel at Samarium's cosmic embrace,
A cosmic traveler, navigating through space.
In its atomic ballet, galaxies align,
Samarium, the element that makes the universe shine.
 So let us cherish Samarium's cosmic reign,
A celestial conductor, orchestrating the plane.

In its atomic symphony, cosmic melodies sound,
Samarium, the element that spreads harmony around.
　　With Samarium as our guide, we explore,
The mysteries of the cosmos, forevermore.
In its atomic embrace, our spirits take flight,
Samarium, the element that illuminates the night.
　　So let us celebrate Samarium's cosmic glow,
A beacon of light in the cosmic flow.
In its atomic dance, the universe sings,
Samarium, the element that grants us cosmic wings.

THIRTY-FOUR

SAMARIUM'S TOUCH

In the realm of elements, a celestial gem,
Samarium, the cosmic conductor, we shall condemn.
In its atomic embrace, it orchestrates the stars,
Samarium, the element that transcends afar.
    Embrace Samarium's essence, a cosmic guide,
Unraveling the secrets that the universe hides.
In its atomic symphony, celestial realms ignite,
Samarium, the element that brings cosmic light.
    Let us cherish Samarium, a cosmic key,
Unlocking the door to the celestial sea.
In its atomic dance, galaxies entwine,
Samarium, the element that makes the cosmos shine.
    With Samarium's touch, the universe unfolds,
Revealing cosmic wonders, untold stories untold.

In its atomic embrace, dreams take flight,
Samarium, the element that paints cosmic delight.

So let us celebrate Samarium's cosmic reign,
A stellar navigator, guiding us through the cosmic domain.
In its atomic symphony, the cosmos sings,
Samarium, the element that gives us cosmic wings.

ABOUT THE AUTHOR

Walter the Educator is one of the pseudonyms for Walter Anderson. Formally educated in Chemistry, Business, and Education, he is an educator, an author, a diverse entrepreneur, and he is the son of a disabled war veteran. "Walter the Educator" shares his time between educating and creating. He holds interests and owns several creative projects that entertain, enlighten, enhance, and educate, hoping to inspire and motivate you.

Follow, find new works, and stay up to date
with Walter the Educator™
at WaltertheEducator.com

www.ingramcontent.com/pod-product-compliance
Lightning Source LLC
LaVergne TN
LVHW052001060526
838201LV00059B/3769